STEM in the MILITARY

BIOSECURITY in Warfare

By Tanner Billings

Cavendish Square

Published in 2023 by Cavendish Square Publishing, LLC
2544 Clinton Street, Buffalo, NY 14224

Website: cavendishsq.com

Library of Congress Cataloging-in-Publication Data

Names: Billings, Tanner, author.
Title: Biosecurity in warfare / Tanner Billings.
Description: Buffalo, New York : Cavendish Square Publishing, [2023] | Series: Stem in the military | Includes index.
Identifiers: LCCN 2022033319 | ISBN 9781502665614 (library binding) | ISBN 9781502665607 (paperback) | ISBN 9781502665621 (ebook)
Subjects: LCSH: Biological warfare--Juvenile literature. | Biological warfare--Prevention--Juvenile literature. | Biological weapons--Juvenile literature. | Bioterrorism--Prevention--Juvenile literature.
Classification: LCC UG447.8 .B425 2023 | DDC 358/.384--dc23/eng/20220712
LC record available at https://lccn.loc.gov/2022033319

Editor: Kate Mikoley
Copyeditor: Michele Suchomel-Casey
Designer: Andrea Davison-Bartolotta

The photographs in this book are used by permission and through the courtesy of: Cover Mihail Siergiejevicz/Shutterstock.com; p. 5 DVIDS/Photo by Petty Officer 2nd Class Jason Poplin/U.S. Navy; pp. 6, 39 courtesy of the National Archives; p. 9 Manfred Ruckszio/Shutterstock.com; p. 10 Jonas Vegele/Shutterstock.com; p. 12 (top) DVIDS/still from Courtesy Video/Medical Research and Development Command; pp. 12 (bottom), 32 DVIDS/Photo by William FBill Discher/Medical Research and Development Command; p. 14 courtesy of CDC.gov; p. 16 courtesy of FBI.gov; p. 18 DC Stock/Alamy Stock Photo; p. 19 Wikimedia Commons/File:Bruce_Ivins_award_ceremony.jpg; p. 22 DVIDS/Photo by Spc. April de Armas/U.S. Army Japan; p. 24 (left) Samvado Gunnar Kossatz/Wikimedia Commons/File:Osho_Drive_By.jpg; p. 24 (right) Samvado Gunnar Kossatz/Wikimedia Commons/File:1983_festival_at_Rajneeshpuram.jpg; p. 27 RosnaniMusa/Shutterstock.com; p. 28 urfin/Shutterstock.com; p. 30 DVIDS/Photo by Lance Cpl. Jerrick J. Griffin/1st Marine Logistics Group; p. 36 DVIDS/Photo by Maj. Heather Bashor/Oregon National Guard Public Affairs Office; p. 41 (chart) Artos/Shutterstock.com; p. 41 (head) E_K/Shutterstock.com; p. 41 (healthcare symbol) Serhii Borodin/Shutterstock.com; p. 41 (map) Aha-Soft/Shutterstock.com; p. 41 (people) Pensiri/Shutterstock.com; p. 41 (beakers) cve iv/Shutterstock.com; p. 42 DVIDS/U.S. Army photo by Staff Sgt. Leia Tascarini.

CPSIA compliance information: Batch #CWCSQ23: For further information, contact Cavendish Square Publishing LLC at 1-877-980-4450.

Printed in the United States of America

CONTENTS

INTRODUCTION

"Biosecurity" is a big word that basically means security, or safety, from harmful organisms, or living things. However, depending on how it's used, the meaning of biosecurity can differ. Different scientific disciplines define it in different ways. For a more specific definition, we can look to the National Academy of Sciences, which defines biosecurity as "security against the inadvertent, inappropriate, or intentional malicious or **malevolent** use of potentially dangerous biological agents or biotechnology, including the development, production, stockpiling, or use of biological weapons as well as outbreaks of newly emergent and epidemic disease."

Terms closely related to biosecurity include "biowarfare," "bioterrorism," and "biodefense."

Biowarfare refers to using biological, or living, toxins or infectious agents—including bacteria, viruses, and fungi—to purposely try to kill or incapacitate people, animals, or plants. This is an act of war and is also sometimes called germ warfare. Bioterrorism is a type of terrorism—the use of violence to scare people as a way to achieve a political goal— that involves the intentional release of biological agents like bacteria or viruses.

Biodefense typically refers to short-term efforts to restore biosecurity to a given group of people in a certain area. Such

Members of the U.S. Navy put on protective gear to stay safe from harmful biological agents.

measures are usually military-driven and aim to help people who have been subjected to biowarfare.

This may all sound like something out of a science-fiction movie, but it is all too real. Today's military must be prepared to respond to bioterrorism attacks at a moment's notice. Luckily, the people on the job are highly trained and ready to act whenever needed.

Throughout history, countries and militaries all over the world have dealt with threats to their biosecurity. As science has evolved, bioterrorists have come up with increasingly harmful ways to attack. The bright side is that countries protecting their people's biosecurity have been able to use advancing science to defend against such threats and attacks.

President Bill Clinton warned of the dangers of biological weapons in his 1998 State of the Union address.

CHAPTER 1

A HISTORY OF BIOWARFARE

On January 27, 1998, President Bill Clinton gave a State of the Union address to the U.S. Congress. In his address, he said, "Together we must also confront the new hazards of chemical and biological weapons and the outlaw states, terrorists, and organized criminals seeking to acquire them."

While President Clinton was right to warn the nation's leaders about the hazards of biological weapons, he was wrong about one thing: They were far from being new hazards. Biological weapons have been around for nearly as long as people have known about infectious disease. While scientific advancements may help bioterrorists come up with new ways to spread the hazards, the weapons themselves are among the oldest still used today.

Early Methods of Bioterror

As early as 600 BCE, armies recognized the potential impact of infectious diseases on opposing armies as well as on civilian populations. Combatants would often pollute wells and other water sources of their enemies with human corpses, animal carcasses, and various types of filth. The Assyrians used rye ergot to poison the wells of their enemies. Ergot is a fungus that grows on rye and produces a chemical similar to the drug LSD. Those who consumed the ergot-infected water became ill or died. The use of such crude methods continued even into the 20th century.

As late as medieval times, people had yet to understand how germs spread disease. From the 1100s through the 1600s, medical professionals believed the foul odor from decaying bodies was the means by which disease was circulated. Still, the actions of militaries were guided by whatever worked—and their coarse attempts at biological warfare worked.

Moving into the 1700s, Russian troops catapulted plague-infected bodies into Swedish cities during the Great Northern War (1710–1721). In addition, British soldiers distributed blankets from smallpox victims to Native Americans in the French and Indian War (1754–1763). In one such incident, according to the Texas Department of State Health Services, smallpox-laden blankets were passed to two members of the Delaware tribe. William Trent, a local militia commander, later noted in his journal, "Out of our regard for them, we gave them two Blankets and a Handkerchief out of the Small Pox Hospital. I hope it will have the desired effect."

In 1797, no less a historical figure than Napoleon Bonaparte himself flooded the plains around Mantua, Italy, in hopes of enhancing the spread of malaria during the siege of an Austrian army post. Later, during the American Civil War (1861–1865), Confederate agents—improvising on a well-tested tactic from earlier wars—sold clothing from yellow fever and smallpox patients to Union troops.

In World War I (1914–1918), Germany shipped infected cattle to the United States and other countries. The Japanese also experimented with biological warfare in occupied Manchuria around 1936.

Rye ergot, which can be seen on this plant, can **contaminate** water, causing serious illness and even death.

Attacking Agriculture?

On September 11, 2001, terrorists crashed planes into the World Trade Center in New York City and the Pentagon in Arlington, Virginia. Another plane, likely intended for a building in Washington, D.C., crashed in a field in Pennsylvania. Since these attacks, experts and ordinary citizens alike have thought about the possibility of a terrorist attack on America's agriculture.

While experts think an agroterrorism event is highly unlikely, they know it's possible. In such an event, a serious disease could shut down the export of every live or processed food or other products at once.

To avoid the possibility of this happening, experts typically recommend stepping up biosecurity. Steps include using gate and door locks for livestock, monitoring livestock for unusual clinical symptoms, screening and training new employees, and watching for strangers or unusual happenings in the area.

Livestock are potential targets of agroterrorism. If they come under attack, it could have a large effect on food availability.

Biowarfare in World War II and Beyond

Numerous incidents of biowarfare occurred during World War II (1939–1945). In China, the Japanese dropped bags of plague-infested fleas on cities. Japanese forces contaminated wells and distributed poisoned foods in China too. Allied scientists experimented with biological agents but did not use them against their enemies.

The United States established the U.S. Army Medical Research Institute of Infectious Diseases (USAMRIID) in 1969 to advance its defenses against biological warfare. Today, the institute continues to work at the cutting edge of the nation's biological defenses.

During the Iran-Iraq War (1980–1988), Iraq's oppressed Kurdish minority sided with the Iranians. On March 17, 1988, the whine of jet engines shattered the predawn stillness in the Iraqi foothills near Iran. A wave of Iraqi Soviet-built MiG fighter-bombers swept across the hills and dropped a combination of antipersonnel cluster bombs and nerve-gas missiles on the Kurdish city of Halabja. The warheads contained various lethal agents, most notably the nerve gases sarin and VX. Twenty MiG bombing runs across the city left around 5,000 Kurds dead and another 20,000 severely wounded.

The atrocity was directed by Iraqi dictator Saddam Hussein. After the Gulf War of 1990–1991, his representatives revealed to members of the UN Special Commissions Team 7 that Iraq had experimented with **anthrax**, food-poisoning bacteria, and botulinum toxins. Saddam's use of chemical warfare, and his admitted attempts to add biological warfare to his **weapons of mass destruction** (WMD) arsenal, would contribute to his end a decade later.

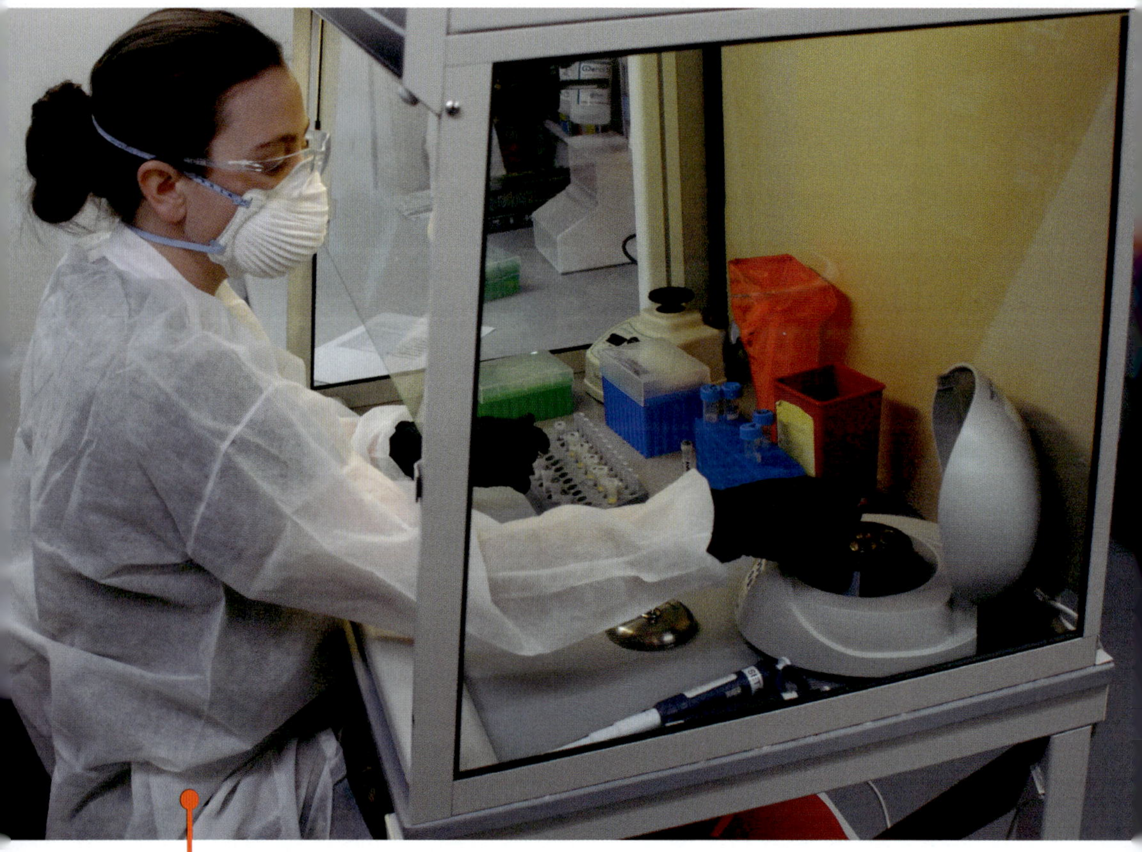

USAMRIID is located at Fort Detrick in Maryland.

Staying on the Defense

The threat of bioterrorism has been a real concern for militaries around the world for centuries, and biodefense is still incredibly important. Having highly skilled teams of experts, such as those at USAMRIID, researching biodefense methods is just one of many important steps in keeping the United States safe from potential biological attacks.

Join the Debate

1. Why do you think biological weapons have been a method of warfare for so long?
2. Why is giving blankets from people with certain infections to others considered biological warfare?
3. Why do you think it's important to be prepared for an agroterrorism event if the possibility of one happening is fairly low?

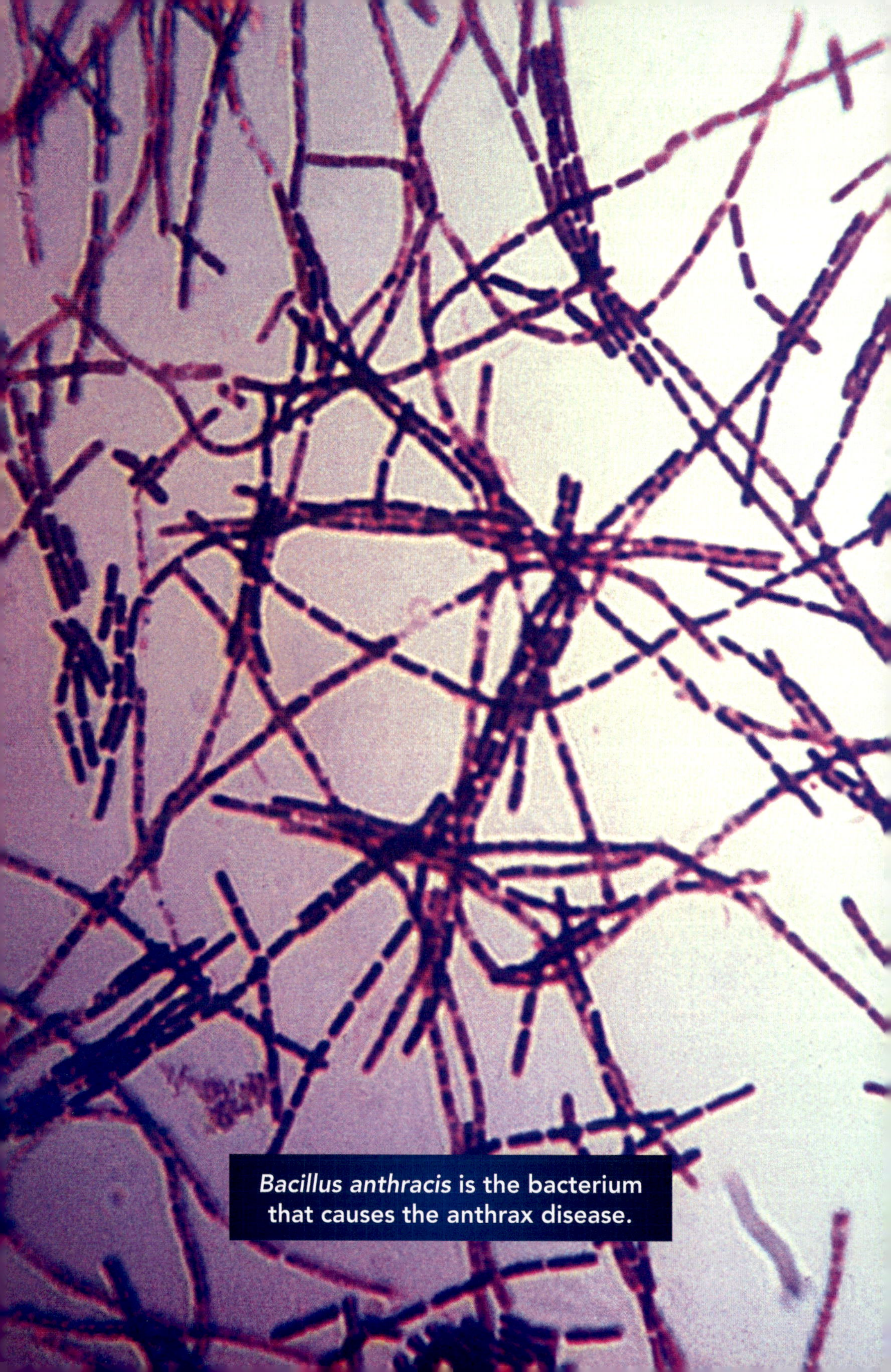

Bacillus anthracis is the bacterium that causes the anthrax disease.

CHAPTER 2

ANTHRAX ATTACKS

Perhaps the most well-known breach of biosecurity in recent history was delivered by mail. The infectious agent was anthrax, a disease transmitted by a bacterium in warm-blooded animals.

On September 18, 2001, exactly one week after the September 11 terrorist attacks, several letters were postmarked from Trenton, New Jersey. Four were sent to the offices of ABC News, CBS News, NBC News, and the *New York Post*, all in New York City. A fifth was sent to the *National Inquirer* at American Media, Inc. (AMI) in Boca Raton, Florida. All contained anthrax **spores**.

The Letters

Only the letters to NBC News and the *New York Post* were recovered. The existence of the other three was presumed because people at ABC, CBS, and AMI became infected. The powdery substance in the *Post* letter was described as a coarse brown granular material.

Three weeks later, on October 9, two more letters were mailed. They bore the same Trenton postmark and contained highly refined powder consisting of almost pure anthrax spores. One was addressed to Democratic senator and Senate Majority Leader Tom Daschle of South Dakota; the other, to Democratic senator and head of the Senate Judiciary Committee Patrick Leahy of Vermont. Neither of the two letters reached their addressees. An aide intercepted the letter to Daschle; Leahy's was rerouted by mistake to the State Department mail annex in Virginia.

The letter to Senator Leahy, shown here, was intercepted and opened safely at Fort Detrick by professionals who wore protective gear.

On October 5, 2001, Robert Stevens of AMI died from inhaling anthrax (called inhalational anthrax). Over the next several weeks, four more people died from inhalational anthrax, and seventeen others were sickened by the anthrax attacks. Most contracted infections were cutaneous (skin) infections.

The Letters' Contents

The letters received at NBC News and the *New York Post* contained photocopies of the same hand-printed threatening note. A similar photocopied note, crudely printed by the same hand, was enclosed in the letters sent to the two senators. Both versions of the note ended with the phrases "Death to America/ Death to Israel/Allah is great." The phrases were meant to suggest a link to the 9/11 terrorists. However, later investigations would show that this was likely a tactic used to try to shift blame from the actual attacker. Regardless, fears of biological terrorism gripped the nation.

The Investigation

Because the U.S. mail was used and state lines were crossed to carry out the crimes, the Federal Bureau of Investigation (FBI) immediately became involved in the investigation. The FBI labeled the case "Amerithrax" and formed the Amerithrax Task Force of close to 30 investigators from several government agencies. The case would go on for the next nine years.

Because of the **sophisticated** refinement of the anthrax spores, the FBI investigation quickly centered on USAMRIID at Fort Detrick, Maryland. During the long, tedious investigation, the FBI eventually turned its attention to Bruce E. Ivins, a biodefense researcher at the institute.

Lab records at Fort Detrick showed that Ivins had logged dozens of late hours before the anthrax letter mailings. Further, a check into his background revealed clear indications of mental instability. For example, in a Department of Justice transcript of a secretly taped interview, Ivins implied he might have done things he could not remember. "I, in my right mind, wouldn't do it," he said of the anthrax attacks. He added, however, "It worries me when I wake up in the morning and I've got all my clothes and my shoes on, and my car keys are right there beside me."

After the letters were discovered, government buildings in Washington, D.C., were regularly checked for anthrax. Anyone who may have come into contact with anthrax needed to wear special protective gear to stay safe.

In July 2008, Ivins learned that the FBI was about to press charges against him in connection with the anthrax attacks. As a possible motive, the *Los Angeles Times* reported that Ivins stood to profit "from massive federal spending in the fear-filled aftermath of those [anthrax] killings." Ivins held patents on anthrax and biodefense vaccines. Before charges were filed against him, Ivins took his own life by overdosing on acetaminophen (Tylenol). He died on August 29, 2008.

The Justice Department, the FBI, and the U.S. Postal Inspection Service announced the formal conclusion of the Amerithrax investigation on February 19, 2010. According to the Justice Department report, "Evidence developed from that investigation established that Dr. Ivins, alone, mailed the anthrax letters."

Some sources criticized the investigation and believed Ivins was innocent, stating there was not enough evidence to prove he was responsible for the letters. However, the case will likely remain closed, and we may never know for certain who sent the letters.

Bruce E. Ivins was a scientist who was developing a vaccine to combat anthrax. The Justice Department suspected Ivins sent the anthrax-laced letters as a way to increase support for the vaccine.

Defending Against Biowarfare

In the late 1990s, scientists from the U.S. Department of Defense (DoD) began to focus on the development and use of vaccines for biodefenses. A 2001 DoD report listed vaccines considered to be "DoD-critical products." Listed were vaccines against anthrax, smallpox, plague, tularemia, botulinum, ricin, and equine encephalitis. The Anthrax Vaccine **Immunization** Program mandates personnel assigned to high-threat areas to receive the anthrax vaccine series.

Defense against biowarfare and bioterrorism targets two distinct populations: civilian noncombatants and military combatants (troops in the field). A critical part of civilian defense requires protection of water sources and food supplies. Defending combat troops against bioweapons relies heavily on protective clothing and gear, as well as vaccinations.

Ramping Up Biosecurity

As tragic as the anthrax panic of 2001 was, some say it resulted in a more aware and better prepared nation. It awakened the country to its vulnerability to biological attacks and encouraged the government and military to work harder to reduce this vulnerability.

In order to strengthen its ability to deal with future investigations, the FBI created the Weapons of Mass Destruction Directorate (WMDD) in July 2006. Its creation, according to an FBI science briefing, "is another example of the FBI's progressive approach focusing on prevention as well as investigation of all issues involving chemical, biological, radiological, and nuclear materials." Part of the WMDD's job is to come up with situations that are worst-case **scenarios**. These are situations we all hope to never have to deal with, but the WMDD imagines these scenarios and comes up with ways to prevent and prepare for them. If these terrible scenarios ever come true, the WMDD will be ready.

Join the Debate

1. Why do you think the anthrax letters caused such a panic around the United States?
2. Do you think the Amerithrax investigation was correct in determining that Dr. Ivins carried out the attacks alone?
3. Would you want to work with the WMDD? Why or why not?

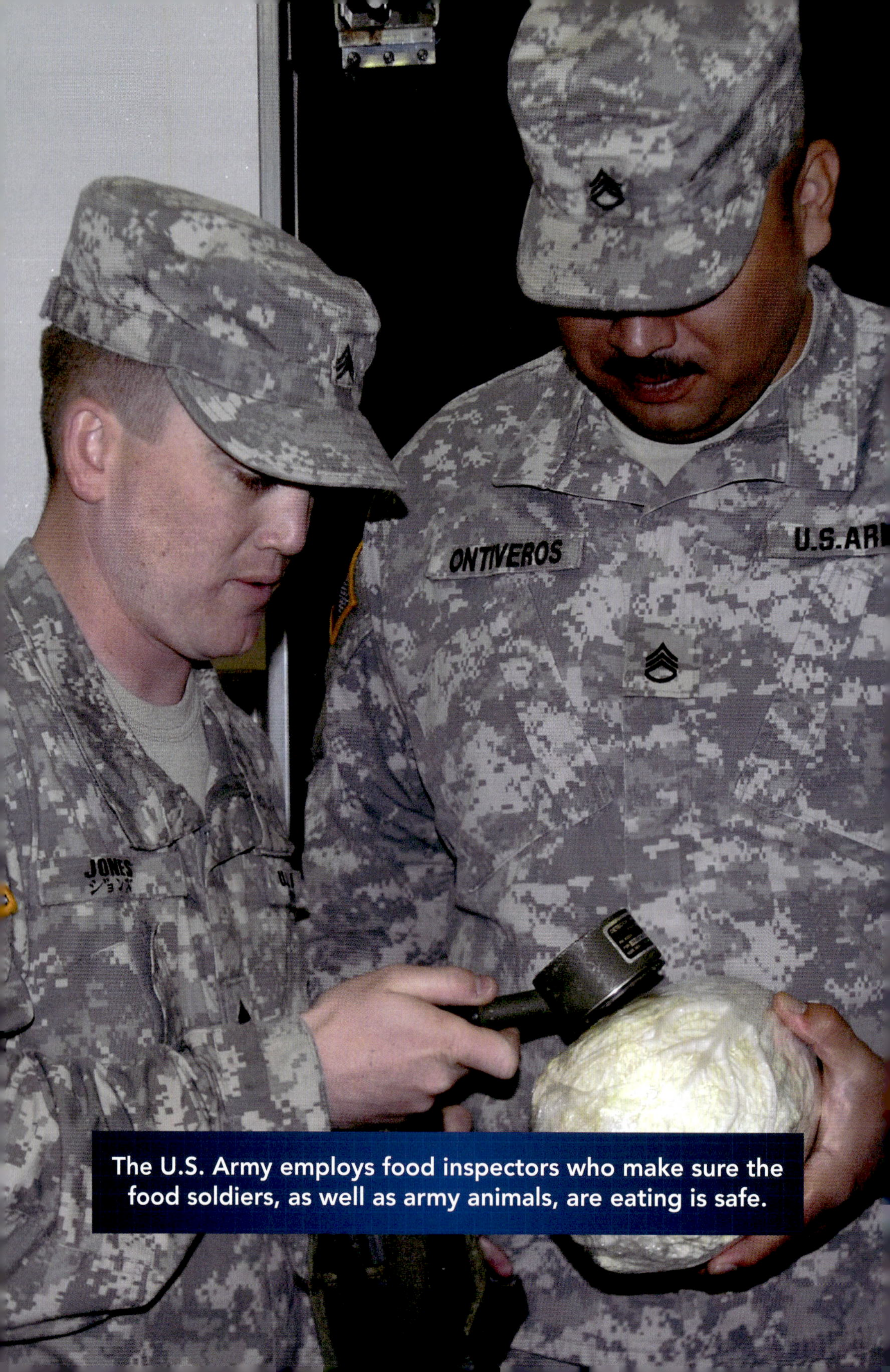

The U.S. Army employs food inspectors who make sure the food soldiers, as well as army animals, are eating is safe.

CHAPTER 3

POISONING FOOD AND WATER SUPPLIES

If you can't count on your food and water being safe, what can you count on? Unfortunately, there have been too many times, even in recent history, when citizens couldn't count on the things they need to survive being safe to consume.

Sometimes this happens accidentally. Food stored in poor conditions can become dangerous to eat. Supplies of drinking water can be mismanaged and inadvertently become poisonous. Other times, however, bioterrorists attack food and water supplies as a means to harm a community.

Danger for Diners

In 1984, the city of The Dalles in Wasco County, Oregon, came under attack. The first call to the Wasco County Health Department came on September 17. A resident of The Dalles reported a case of food poisoning after dining at a restaurant in town. It was the first of many such calls from diners in local restaurants. Within a week of the outbreak, the Centers for Disease Control and Prevention (CDC) determined the cause as *Salmonella typhimurium*. The bacterium eventually infected around 750 people in the town of 10,000.

In the early 1980s, a gray-bearded guru calling himself Bhagwan Shree Rajneesh arrived in New York City from India. (*Bhagwan* is Hindi for "god.") He soon migrated to Oregon and bought a 100-square-mile (259-square-kilometer) site once known as the Big Muddy Ranch. There, he formed a commune of followers. Dressed in red and wearing beaded necklaces with pictures of their leader, they became known as Rajneeshees.

Bhagwan Shree Rajneesh is shown here driving past some of his followers in their community known as Rajneeshpuram.

At first, citizens of the nearby town of Antelope accepted their presence with a live-and-let-live attitude. Neighbors grew concerned, however, as they watched the Rajneeshees build a small city in an area that was legally designated as farmland. The burgeoning city included not only roads, dairy barns, and greenhouses, but also cafeterias, hotels, and medical clinics. Local concerns turned to alarm when the Rajneeshees began attending town council meetings in nearby Antelope in an effort to gain political power.

Determined to seize political control of Wasco County, the cult eventually turned to germ warfare in an effort to keep opposing voters out of the polling booths. They planned to contaminate the county water supply with *Salmonella* bacteria cultivated in their own medical laboratories. To test its potency, they laced the salad bars in The Dalles with their homemade germs. One restaurant owner later said he lost $465,000 in sales and liability claims from infected customers. During the following months, citizens of the county lived in fear over concerns that cult members might poison the water or spread other illnesses.

Fortunately for the people of Wasco County, the Rajneesh cult collapsed under federal investigation and prosecution in 1985. Some cult members were imprisoned; others were deported. Rajneesh himself was fined $400,000 for immigration fraud and was deported—and the largest attack of bioterrorism on U.S. soil to date ended.

Avoiding Attacks on the Army

Imagine what would happen if an army's enemy contaminated its food supply. At the very least, it would harm the army's ability to operate. That is why protection of the army's food and water supplies is a major function of the U.S. Army Public Health Command (USAPHC).

"The primary focus of food defense is the prevention of the intentional contamination of our food supply, while food safety is aimed at preventing unintentional or accidental contamination," explained Sgt. 1st Class Kevin M. Gill, in an article by Jane Gervasoni on the army.mil website.

"Although food safety had been a mainstay of Army veterinary food inspectors, food defense had not even been a focus of discussion until after September 11, 2001," added USAPHC

Attacks on Water?

A Department of Homeland Security (DHS) bulletin issued in 2011 stated, "Over the last decade, violent extremists have expressed aspirational interest in contaminating unspecified water supplies, and as recently as July 2011 specifically raised backpressure as a means of contamination."

"Backpressure happens when there is a pressure, higher than the city water pressure, exerted on the water systems and it causes the water to flow backward," explained Roy Dillard, in an article on the DHS website.

Backpressure commonly occurs when a water main breaks. Anything in that line would get **siphoned** into the main line. Using backpressure, terrorists might introduce chemical or biological agents into the water supply and spread them over large distances long before their detection.

"It's possible," Dillard said. "Virtually any kid taking a high school science class could figure it out, but it would have to be done in such volumes that it would be relatively hard to do because ... we use backflow prevention assemblies in major areas and in almost all of your industrial areas."

All 50 states require controlled cross-connections and backflow prevention systems. Even if terrorists tried to inject a biological agent into a water supply, chlorine in the tap water would likely nullify it. "Our water systems are pretty well protected," Dillard said.

Backpressure can occur after a water main pipe breaks.

Veterinary Services Food Protection Program manager Col. Thomas E. Honadel. "Since that time, more emphasis has been placed on antiterrorism food defense plans."

"Food defense measures include training for food service personnel, increased security of food service areas, and even background and identification checks," Gill continued. Personnel like Gill and Honadel protect the food supplies of the U.S. Army.

Breaking Down Bioterrorism Agents

Salmonella, the type of bacteria used in the attacks on Wasco County, are harmful bacteria that commonly cause food poisoning. According to the U.S. Department of Health and Human Services (HHS), *Salmonella* bacteria cause 1.2 million illnesses, 23,000 hospitalizations, and 450 deaths each year. The CDC separates *Salmonella* and other bioterrorism agents into three categories:

Category A organisms or toxins pose the greatest public health and national security risks: They can easily be spread from person to person; result in high death rates; might cause major health impacts, public panic, and disruption; and require special action for health preparedness.

Category B agents are fairly easy to spread, result in moderate illness rates and low death rates, and need special laboratory enhancements and enhanced disease monitoring.

Category C includes new pathogens (infectious agents) that might later be prepared for mass spread. They are readily available, are easy to produce, and have the potential for major health impact, including high disease and death rates.

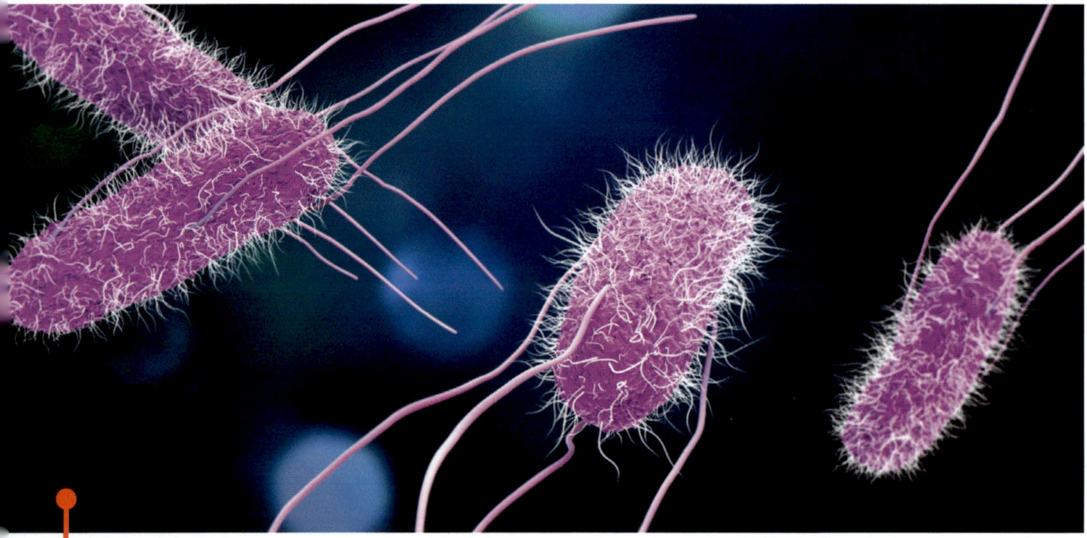

Salmonella bacteria, depicted here, are a CDC Category B agent.

Before and during the Gulf War, coalition forces anticipated the enemy's potential use of biological and chemical weapons. They trained in protective masks and equipment, practiced decontamination procedures, and received extensive instructions on detection and immunization methods. About 150,000 U.S. troops were vaccinated against anthrax, and another 8,000 received a new botulinum vaccine.

In the civil defense sector, experts recognize biosurveillance as the single most important tool for identifying public health events of worldwide significance. Emerging infectious diseases reside high on the list.

Biosurveillance professionals perform several basic functions. The first is to detect and report disease within a given population. They must also analyze and confirm reported cases to detect outbreaks and be ready to respond quickly and appropriately at the local or regional levels to prevent and control disease outbreaks at the national level. Finally, these experts need to provide information on the causes, distribution, and control of diseases in populations. With biosurveillance systems in place, the United States makes sure there are experts on constant alert prepared to protect the biosecurity of all Americans.

Join the Debate

1. How do you think the Wasco County attacks impacted restaurants in the area?
2. Why do you think it's especially important for the army and other armed forces to inspect service members' food?
3. Which biosurveillance functions do you think are most important?

Chemical, biological, radiological, and nuclear defense specialists in the U.S. Marines take part in a practice decontamination exercise.

CHAPTER 4

FIGHTING BIOLOGICAL THREATS

A bioterrorist attack may seem unlikely, but unfortunately there are various terrorist organizations around the world that hold no qualms about carrying out massive deadly attacks on people who disagree with their beliefs. Their existence demands that all free nations mount a constant and vigorous biodefense. USAMRIID stands at the forefront of America's defense against bioterrorists.

All About USAMRIID

USAMRIID specializes in research on biological threats. Its primary job is to provide medical solutions for the protection of service members, but its expertise can also be used to protect the entire population as needed by developing medical solutions—vaccines, drugs, diagnostics, and information—to bioterror threats. It is the only Department of Defense laboratory equipped to safely study highly hazardous infectious agents that require maximum containment at Biosafety Level 4 (BSL-4). A biosafety level is the level of containment needed to work with a given biological agent. It is based on the risk posed by the agent and the activities involved with the testing. Scientists at USAMRIID strive every day to maintain the laboratory's world-class scientific and technology base.

USAMRIID is a national laboratory. It, and other laboratories like it, represent the key elements of the Laboratory Response Network (LRN). In addition to its primary work to protect military troops, USAMRIID's scientists provide expert consultation and training for medical personnel. LRN laboratories are responsible for characterizing specialized strains, **bioforensics**, and handling highly infectious biological agents. Such work requires much planning, as well as state-of-the-art facilities for the safety and security of the scientists.

USAMRIID has played a critical role in attempting to combat the COVID-19 pandemic. Here, a worker is harvesting samples of the virus intended to be used to make models of infection, diagnostic tests, and vaccines.

At USAMRIID, most LRN testing is done in laboratories equipped for Biosafety Level 3, but it is one of only a few U.S. laboratories with BSL-4 facilities. Each biosafety level requires its own set of safety equipment, facilities, and practices to reduce the risk of getting infected in the laboratory. Personnel working in BSL-4 laboratories, for example, must wear full-body pressurized suits that resemble the space suits worn by astronauts. The suits provide the highest form of personal protective equipment. Those who work in BSL-4 labs must also take chemical showers, then enter a vacuum room to be exposed to ultraviolet light. This process destroys any trace amounts of the organisms after testing.

USAMRIID AREAS OF EXPERTISE

- antiviral drug protocols
- applied medical research
- biodefense
- biological threat research
- biosafety
- containment
- diagnosis and treatment
- disease identification
- hazardous infectious agents
- management of biological casualties
- outbreak assessment and control
- vaccines
- virus analysis

USAMRIID is the lead medical research laboratory for the U.S. Biological Defense Research Program. These are its areas of expertise.

Biowarfare Training

The training of worldwide first responders is high on the list of the services provided by USAMRIID. First responders include hazardous materials teams, National Guard Support Teams, tactical response teams, emergency communications, and others. USAMRIID's Field Operations and Training (FO&T) Branch offers two first-responder courses: Biological Agent Identification and Counterterrorism Training (BAIT) and Field Identification of Biological Warfare Agents (FIBWA).

The FIBWA program consists of four different core courses: Basic Course, Manager's Course, National Guard Bureau & Civil Support Teams, and Special Interest Training. The program runs for 20 working days. Students learn to set up, maintain, and operate mobile laboratories under field conditions for both biowarfare and biosurveillance missions. Emphasis is placed on using a combination of several technologies. These lessons are coupled with hands-on laboratory experience aimed at reaching a highly reliable solution.

BAIT

Biological Agent Identification and Counterterrorism Training (BAIT) teaches responders to be prepared for biological threats. Experts stage customized scenarios for single organizations or several at once. The staff then monitors and reviews training operations with responders after completion. According to USAMRIID's website, BAIT provides realistic training for those responding to bioterrorism events, true subject matter experts for recommending improvements to organizational functions, custom training experience based on each specific unit's needs, and reinforcement of positive relationships between organizations.

America's Biodefense Vanguard

USAMRIID stands at the vanguard of America's defense against a full range of biological threats—both on the battlefield and on the domestic front. The skill and experience of those employed by USAMRIID to combat the most dangerous possible biological attacks is second to none. This team works behind the scenes

Gaming on the Job?

Personnel at the U.S. Army Edgewood Chemical Biological Center (ECBC) are working hard to help soldiers learn to combat chemical biological attacks. ECBC, part of the U.S. Army Research, Development, and Engineering Command, employs a special staff to develop multimedia and interactive training aids for soldiers. These team members—including engineers, product developers, and animators—typically have professional backgrounds in computer science and art. They use their skills to create realistic training programs to train soldiers on Chemical, Biological, Radiological, Nuclear, and Explosives (CBRNE) missions. To create these trainings, they use video game technology and virtual reality headsets, making for training methods that are fun and effective!

to ensure the country is always prepared to take on the types of bioterror attacks we hope to never face.

Join the Debate

1. Why do you think it's important that people studying biological threats wear protective gear?
2. Why do you think we need highly skilled first responders prepared for biowarfare attacks?
3. How important do you think the training materials developed by the team at ECBC are for soldiers learning to combat biological attacks?

Members of the Army National Guard helped distribute COVID-19 vaccines to the public.

CHAPTER 5

TODAY AND TOMORROW

In the mid-1300s, bubonic plague killed 60 percent of the European population. The plague, commonly known as the Black Death, caused huge boils on its victims, who then suffered a slow, painful death.

In 2019, the COVID-19 pandemic began. It soon took over the world. By July 2022, more than a million people had died in the United States alone from the harmful virus.

In today's world, it's easy to wonder what would happen if terrorists got their hands on a harmful bacteria or virus to use as a weapon of biowarfare. With the prominence of certain terrorist groups around the world, some experts worry such an attack is a possibility. That's why the work of organizations such as USAMRIID is so important and must continue.

Intent to Use Viral Weapons

In an August 2014 article in *Foreign Policy*, Harald Doornbos and Jenan Moussa detailed evidence of deadly terrorist intent. A laptop computer belonging to an ISIS terrorist named Muhammed S. was found in Syria. It contained a

19-page primer in Arabic on developing biological weapons and weaponizing bubonic plague from infectious animals. Translated, it said in part, "The advantage of biological weapons is that they do not cost a lot of money, while human casualties can be huge."

The document went on with instructions on how to achieve the best results with a viral weapon. A translation read, "Use small grenades with the virus, and throw them in closed areas like metros [subways], soccer stadiums, or entertainment centers. Best to do it next to air-conditioning." It left little doubt as to the evil aspirations of the terrorists.

Preventing Biowarfare

For most people, few things can be more frightening than the intentional contamination of life-sustaining necessities—air, water, or food supplies—by bioterrorists. World leaders recognized the dangers of germ warfare shortly after World War I. In 1925, they signed the Geneva Protocol prohibiting the use of chemical and biological weapons in war. A half-century later, the provisions of the Geneva Protocol were supplemented by those of the 1972 Biological and Toxin Weapons Convention (BWC).

The BWC took effect in 1975. By the turn of the century, around 144 nations had signed onto it. However, additional provisions and legislation that have since been proposed to control bioweapons have often received pushback, including from the United States. American leaders have at times expressed fear that some new proposals might reveal vital U.S. military secrets and thus pose a threat to national security.

Bioweapons Convention

The 1972 Biological and Toxin Weapons Convention (BWC) was the first international treaty to ban an entire class of weapons. Article I of the BWC, as noted by the Federation of American Scientists, reads as follows:

> *Each state party to this Convention undertakes never in any circumstances to develop, produce, stockpile or otherwise retain: (1) Microbial or other biological agents, or toxins whatever their origin or method of production, of types and in quantities that have no justification for prophylactic [disease preventative], protective or other peaceful purposes; (2) Weapons, equipment or means of delivery designed to use such agents or toxins for hostile purposes or armed conflict.*

The BWC treaty includes additional articles to ensure continued international cooperation and agreed-upon implementation of its provisions.

President Richard Nixon is shown here with other representatives at the Biological and Toxin Weapons Convention.

Failure of safeguards to keep pace with biological advances begs the question of how the world can defend against evolving bioweapons.

Many methods are needed to defend against existing and evolving forms of bioweapons. High on the list is continual improvement of and stricter adherence to existing international conventions banning WMDs. Some militants suggest the use of nuclear retaliation to deter bioterrorists. Perhaps the best biodefense resides with emerging tools of genetic knowledge and biological technology. These include immune system enhancement, new vaccines, antibiotics and antiviral drugs, and efficient bioagent detection and identification equipment. Most importantly for Americans, the U.S. government needs to actively pursue methods of biodefense to protect its citizens.

That's why in 2018, the U.S. government unveiled its National Biodefense Strategy (NBS). The strategy intended to address challenges stemming from "naturally occurring, deliberate, or accidental biological threats." The U.S. Department of Health and Human Services, which oversees the NBS, stated that throughout the COVID-19 pandemic, the Biodefense Coordination Team continued to implement the strategy in an effort to protect the nation from both present and possible future biological threats.

As of 2022, 17 U.S. departments and agencies take part in biodefense efforts. While biological threats are nothing new, today's efforts to protect against these threats are stronger than ever.

This infographic shows the five main objectives of the National Biodefense Strategy.

The Tennessee Army National Guard takes part in training to identify chemical, biological, radiological, nuclear, and explosive agents.

Join the Debate

1. While the COVID-19 pandemic was not a result of bioterrorism, it made many people think about how their country would respond to a biological attack. Do you think how the government responded to the pandemic accurately represents how it would respond to an act of bioterrorism?
2. Why do you think the BWC is important?
3. How do you think the NBS contributes to biosafety in the United States?

GLOSSARY

anthrax: A disease that affects animals, such as cattle, and sometimes people. It features symptoms that range from skin ulcers to breathing problems.

bioforensics: A field of microbiology dealing with biological agents or their components.

contaminate: To pollute.

immunization: The act of making a human or animal immune, or resistant, to infection.

malevolent: Something productive of harm or evil.

scenario: A sequence of events; an account or synopsis of a possible course of action or events.

siphon: To move a liquid from one container or area to another.

sophisticated: Highly complicated or developed.

spore: A reproductive body produced by fungi and some plants and bacteria that consists of a single cell able to produce a new individual.

weapon of mass destruction: A chemical, biological, or radiological weapon capable of causing widespread death and destruction.

FIND OUT MORE

Books

Beales, Paul, John Elphinstone, Adrian Fox, Charles Lane, Derek McCann, Tim Lacey, Julian Little, Kerry Maguire, and Alice Turnbull. *Plant Diseases and Biosecurity*. Oxford, United Kingdom: Oxford University Press, 2019.

Collins, Anna. *Biological Weapons: Using Nature to Kill*. New York, NY: Lucent Press, 2018.

Websites

Biosecurity
www.phe.gov/s3/BioriskManagement/biosecurity/Pages/default.aspx
The Science Safety Security project is a collaboration of several U.S. government departments aiming to educate about biorisk management. This page has information on biosecurity.

Bioterrorism
www.interpol.int/en/Crimes/Terrorism/Bioterrorism
This page on the website of the International Criminal Police Organization (INTERPOL) has lots of information on bioterrorism.

Organizations

Centers for Disease Control and Prevention (CDC)
1600 Clifton Road
Atlanta, GA 30329
www.cdc.gov
The CDC is a science-based government organization that protects the American public's health. The organization's website has information on bioterrorism and other hazards.

Ready Campaign
FEMA/DHS
500 C Street SW
Washington, DC 20472
www.ready.gov
The Ready Campaign promotes preparedness for emergencies, including natural and man-made disasters. There is a whole section on its website about bioterrorism.

U.S. Army Medical Research Institute of Infectious Diseases (USAMRIID)
1425 Porter Street
Frederick, MD 21702-5011
www.usamriid.army.mil
USAMRIID is the military's lead laboratory for medical biodefense research.

Publisher's note to educators and parents: Our editors have carefully reviewed these websites to ensure that they are suitable for students. Many websites change frequently, however, and we cannot guarantee that a site's future contents will continue to meet our high standards of quality and educational value. Be advised that students should be closely supervised whenever they access the internet.

INDEX

ABOUT THE AUTHOR

Tanner Billings is a writer living in New York. He enjoys studying the history of the military and keeping up with modern military technologies. When he is not writing, he enjoys hiking and taking trips to national parks with his family and dogs.